Tucholsky Wagner Zola Scott Sydow Freud Schlegel
Turgenev Wallace Fonatne
Twain Walther von der Vogelweide Fouqué Friedrich II. von Preußen
Weber Freiligrath Frey
Fechner Fichte Weiße Rose von Fallersleben Kant Ernst Richthofen Frommel
Hölderlin
Engels Fielding Eichendorff Tacitus Dumas
Fehrs Faber Flaubert
Eliasberg Ebner Eschenbach
Feuerbach Maximilian I. von Habsburg Fock Eliot Zweig
Ewald Vergil
Goethe Elisabeth von Österreich London
Mendelssohn Balzac Shakespeare Dostojewski Ganghofer
Trackl Lichtenberg Rathenau Doyle Gjellerup
Mommsen Stevenson Tolstoi Hambruch
Thoma Lenz Hanrieder Droste-Hülshoff
Dach Verne von Arnim Hägele Hauff Humboldt
Karrillon Reuter Rousseau Hagen Hauptmann Gautier
Garschin Defoe Hebbel Baudelaire
Damaschke Descartes
Hegel Kussmaul Herder
Wolfram von Eschenbach Darwin Dickens Schopenhauer Rilke George
Bronner Melville Grimm Jerome Bebel
Campe Horváth Aristoteles Barlach Proust
Bismarck Vigny Voltaire Federer Herodot
Gengenbach Heine
Storm Casanova Tersteegen Gilm Grillparzer Georgy
Chamberlain Lessing Langbein Gryphius
Brentano Lafontaine
Strachwitz Claudius Schiller Kralik Iffland Sokrates
Bellamy Schilling
Katharina II. von Rußland Gerstäcker Raabe Gibbon Tschechow
Löns Hesse Hoffmann Gogol Wilde Gleim Vulpius
Luther Heym Hofmannsthal Klee Hölty Morgenstern Goedicke
Roth Heyse Klopstock Kleist
Luxemburg Puschkin Homer Mörike
La Roche Horaz Musil
Machiavelli Kierkegaard Kraft Kraus
Navarra Aurel Musset Kind Moltke
Nestroy Marie de France Lamprecht Kirchhoff Hugo
Laotse Ipsen Liebknecht
Nietzsche Nansen
Marx Lassalle Gorki Klett Ringelnatz
von Ossietzky May Leibniz
vom Stein Lawrence Irving
Petalozzi Knigge
Platon Pückler Michelangelo Kock Kafka
Sachs Poe Liebermann
de Sade Praetorius Mistral Zetkin Korolenko

The publishing house tredition has created the series **TREDITION CLASSICS**. It contains classical literature works from over two thousand years. Most of these titles have been out of print and off the bookstore shelves for decades.

The book series is intended to preserve the cultural legacy and to promote the timeless works of classical literature. As a reader of a **TREDITION CLASSICS** book, the reader supports the mission to save many of the amazing works of world literature from oblivion.

The symbol of **TREDITION CLASSICS** is Johannes Gutenberg (1400 – 1468), the inventor of movable type printing.

With the series, tredition intends to make thousands of international literature classics available in printed format again – worldwide.

All books are available at book retailers worldwide in paperback and in hardcover. For more information please visit: www.tredition.com

tredition was established in 2006 by Sandra Latusseck and Soenke Schulz. Based in Hamburg, Germany, tredition offers publishing solutions to authors and publishing houses, combined with worldwide distribution of printed and digital book content. tredition is uniquely positioned to enable authors and publishing houses to create books on their own terms and without conventional manufacturing risks.

For more information please visit: www.tredition.com

The Peanut Plant Its Cultivation And Uses

B. W. Jones

Imprint

This book is part of the TREDITION CLASSICS series.

Author: B. W. Jones
Cover design: toepferschumann, Berlin (Germany)

Publisher: tredition GmbH, Hamburg (Germany)
ISBN: 978-3-8495-0523-3

www.tredition.com
www.tredition.de

Copyright:
The content of this book is sourced from the public domain.

The intention of the TREDITION CLASSICS series is to make world literature in the public domain available in printed format. Literary enthusiasts and organizations worldwide have scanned and digitally edited the original texts. tradition has subsequently formatted and redesigned the content into a modern reading layout. Therefore, we cannot guarantee the exact reproduction of the original format of a particular historic edition. Please also note that no modifications have been made to the spelling, therefore it may differ from the orthography used today.

PREFACE.

This little work has been prepared mainly for those who have no practical acquaintance with the cultivation of the Peanut. Its directions, therefore, are intended for the beginner, and are such as will enable any intelligent person who has followed farming, to raise good crops of Peanuts, although he may have never before seen the growing plant.

The writer has confined himself to a recital of the more important details, leaving the minor points to be discovered by the farmer himself. If the reader should think these pages devoid of vivacity, let him remember that we have treated of an every-day subject in an every-day style. The interest in the theme will increase when the beginner has pocketed the returns from his first year's crop. Until then, we leave him to plod his way through the details, trusting that the great Giver of the harvest will bless his labors, and amply reward his toils in this new field.

B. W. J.

Warren Place, Surry County, Va., 1885.

CONTENTS.

CHAPTER I. — Description.

Origin. — Natural History. — Varieties. — Possible Range. — Analysis.

CHAPTER II. — Planting.

Soil, and Mode of Preparation. — Seed. — Time and Mode of Planting. — Fertilizers. — Replanting. — Moles, and Other Depredators. — Critical Period.

CHAPTER III. — Cultivation.

First Plowing and Weeding. — Subsequent Workings. — Implements. —
When Cultivation should Cease. — Insect Enemies. — Effects of Cold.
— Effects of Drouth. — Appearance at this Period.

CHAPTER IV. — Harvesting.

When to begin Harvesting. — Mode of Harvesting. — Why cured in the Field. — Depredators. — Detached Peanuts. — Saving Seed Peanuts.

CHAPTER V. — Marketing.

Picking the Peanuts. — Price paid Pickers. — Cleaning and Bagging. — Peanut "Factories." — The best Markets. — Picking Machines.

CHAPTER VI. — Uses.

Peanut Oil. — Roasted Peanuts. — Peanut Candy. — Peanut Coffee. — Peanut Chocolate. — Peanut Bread. — Peanut Soap. — Peanuts as a Food for Stock. — Peanut Hay.

APPENDIX.

A. Statistics.

B. Costs.

C. The Peanut Garden of America.

THE PEANUT PLANT;

ITS CULTIVATION AND USES.

CHAPTER I.

DESCRIPTION.

Origin.—The native country of the Peanut (*Arachis hypogæa*) is not definitely ascertained. Like many other extensively cultivated plants, it has not been found in a truly wild state. Some botanists regard the plant as a native of Africa, and brought to the New World soon after its discovery. Sloane, in his history of Jamaica, states that peanuts formed a part of the provisions taken by the slave ships for the support of the negroes on the voyage, and leaves it to be inferred that the plant was introduced in this manner. De Candolle, in *Géographie Botanique Raisonnée*, and his latter work on *L'Origine des Plantes Cultivées*, strongly inclines to the American origin of the Peanut. The absence of any mention of the plant by early Egyptian and Arabic writers, and the fact that there is no name for it in Sanscrit and Bengalese, are regarded as telling against its Oriental origin. Moreover, there are six other species of *Arachis*, natives of [Pg 6] Brazil, and Bentham and Hooker, in their *Genera Plantarum*, ask if the plant so generally grown in warm countries may not be a cultivated form of a Brazilian species.

If, as seems probable, the Peanut is really a native of America, then this Continent has contributed to the agricultural world five plants that have exerted, and will continue to exert, an immense influence on the industries and commerce of the world. These are: the Potato, Cotton, Tobacco, Indian Corn, and the Peanut. Of these five, the Peanut, the last to come into general and prominent notice, is destined to rival some of the others in importance.

Whatever may have been its origin, the Peanut plant has gradually made its way over an extended area of the warmer parts of both the Old and New World, and in North America has gained a permanent foot-hold in the soil of the South Atlantic and Gulf States. Nor has it yet reached its ultimate limits, for cultivation and acclimation will inure it to a sterner climate, until it becomes an important crop in latitudes considerably further north than Virginia. This is indicated by its rapid spread within the past few years. Remaining long in comparative obscurity, it was not until a recent period that the Peanut gained prominence as an agricultural and commercial staple, but since it fairly started, its progress has been rapid and sure.

Natural History.—There are some peculiarities about the Peanut plant that make it interesting to the naturalist. Its habit of clinging close to the soil, the closing [Pg 7] together of the leaves at sunset, or on the approach of a storm, the beautiful appearance of a field of it when full grown, and the remarkable wart-like excrescences found upon the roots, are some of its more notable characteristics. Its striking preference for a calcareous soil is another of its peculiarities, the Peanut producing more and better crops on this kind of soil than on any other.

The Peanut belongs to the Natural Order *Leguminosæ*, or pod-bearing plants, and this particular member of it is as unlike all the rest with which we are acquainted, as can well be conceived. No other grows so recumbent upon the soil, and none but this produces seed under ground.

The botanical name of the Peanut is *Arachis hypogæa*. The origin of the generic name *arachis* is somewhat obscure; it is said to come from *a*, privative, and *rachis*, a branch, meaning having no branches, which is not true of this plant. The specific appellation, *hypogæa*, or "under-ground," describes the manner in which the pods grow. The following is a partially technical description of the plant:

Root annual, branched, but not fibrous, yellowish, bitter, and warty; Stem procumbent, spreading, much-branched, somewhat hairy towards the extremities; Leaves compound, leaflets obovate, mucronate, margin entire, ciliate when young, smooth and almost leathery with age, leaves closing at night and in rainy weather; Flowers papilionaceous, yellow, borne upon the end of an axillary peduncle. After flowering, the forming-pod is, by the elongation of its stalk, pushed into the soil, beneath which it grows and ripens; Legume, or pod [Pg 8] indehiscent, woody and veiny, one to four-seeded; Seed, with a reddish coat, the embryo with two large, fleshy cotyledons, and a very short, nearly straight, radicle. Figure 1 represents a portion of the Peanut plant.

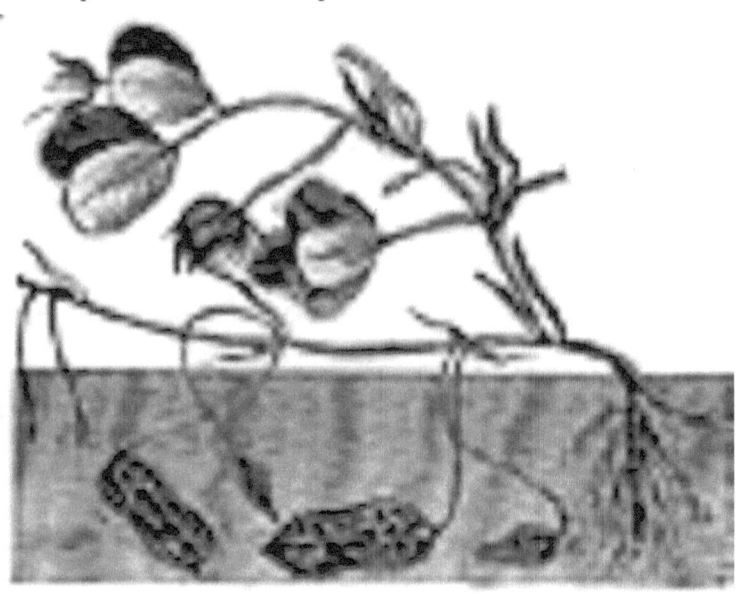

Fig. 1.—PORTION OF THE PEANUT PLANT, showing how the minute pods from above-ground flowers are forced into the soil to grow and ripen.

Varieties.—While no botanical varieties of *Arachis hypogæa* have been described, its long cultivation in different countries in unlike soils and climates, has produced several cultural varieties. Taking the Virginia Peanut as the typical form, there may be named as differing from it, the North Carolina Peanut, having very small but solid and heavy pods, that weigh twenty-eight [Pg 9] pounds to the bushel. The Tennessee Peanut is about the size of the Virginia variety, but has a seed of a much redder color and less agreeable flavor. There is a Bunch variety, that does not spread out like a mat over the soil, but grows upright like the common field pea. This last kind has been raised to some extent in Virginia, but has never become popular with planters, and is fast passing out of cultivation. It is possible that the Bunch Peanut is a representative of the plant in its wild state. It produces fewer seeds and less vine than any other kind. The flat or spreading Peanut shows a tendency to sport in this direction, and in any large field of peanuts, quite a number of plants will be found that have the bunch form, and such are always barren or seedless hills.

The small-podded, or North Carolina Peanut, is not at all popular with pickers, because it takes a great many more to make a basketful, and, unless they are paid an extra price for picking this sort, they cannot make as good wages. Nor do our planters seem to like it very well, finding it more trouble to handle than the larger variety. Hence it is but little cultivated in Virginia.

The Peanut in its travels has also acquired a variety of names, such as ground-pea, earth-nut, goober [1] or guber, and pindar. Also "currency," "cash," "credit," and other expressive titles. Of all these names, "Peanut" [Pg 10] is the most generally used, but Ground-pea would be the more descriptive name.

Possible Range.—From a somewhat careful study of the climatic requirements of the Peanut plant, and of the isotherms of summer temperature, we are satisfied that it would thrive as far north as the northern limit of the zone of the vine. This for the United States, as delineated in Mitchell's Physical Geography, starts on the Pacific Coast in the latitude of British Columbia, turns suddenly south along the Cordilleras to Colorado, then trends as suddenly northward to the northern limits of Iowa, strikes eastwardly along a line to the south of the great lakes, and enters the Atlantic in the vicinity of Cape Cod. If our view is correct, the Peanut will thrive on any suitable soil within the limits of the United States lying to the south of this line. This would make the cultivation of the Peanut possible in by far the greater part of the entire country. In fact, there is no doubt but that it may be grown successfully wherever Indian corn will thrive luxuriantly. Any section having a growing season of five months exempt from frost, may raise the Peanut. This gives the crop a much wider range than has been thought possible. It does not require a long period of extreme heat to mature it. The seeds are mostly formed in the cooler weather of the latter part of summer and the first of autumn. Planted in June, cultivated until August or a little later, and harvested the last of September, it can be perfected in four months, though the Virginia planter takes five months for it. Any good calcareous soil, west of New [Pg 11] Jersey and southward, that is not too elevated, will grow the Peanut.

Analysis.—This, perhaps, is not a matter of much practical importance to the planter. The best peanut soil and the proper fertilizer had been found out before an analysis of the plant had been made. Still there are some advantages in knowing what are the prominent elements that enter into the composition of this, or any other, cultivated plant, and an analysis is accordingly given.

An analysis made by Doctor Thomas Antisell, chemist to the Department of Agriculture at Washington, and published in the Report of that Department about the year 1869, gives the following as the composition of the Peanut plant:

In one hundred parts of the husk and nut taken together

Water	2.60
Albuminous, fibrous matter and starch	79.26
Oil	16.00
Ash	2.00
Loss	.14
	100.00

In one hundred parts of the husk and seed separated:

	Seed.	*Husk.*
Moisture	2.51	2.61

Albuminous matter and farina	79.71	traces.
Cellulose		85.48
Ash	1.77	11.90
Oil	16.00	
	99.99	99.99

[Pg 12] "The ash of the seed," it was stated by the same authority, "consists of salts wholly soluble in water, composed of the phosphates of alkalies, with traces of alkaline, chlorides, and sulphates. The ash of the husk differs, in consisting chiefly of common salt, phosphate of lime and magnesia."

The analysis of the ash of the Peanut, furnished to the *American Agriculturist*, by H. B. Cornwall, Professor of Analytical Chemistry in the John C. Green School of Science, College of New Jersey, Princeton, and published in that Journal for July, 1880, gives the following as the mineral elements of this plant:

PER ONE HUNDRED PARTS OF ASH.

Silica	1.06
Potash	44.73
Soda	14.60
Lime	1.71
Magnesia	12.65
Phosphoric acid	17.64
Sulphuric acid	2.53
Chlorine	0.15
	95.07

In this analysis neither the carbonic acid nor carbon were determined.

It was further stated that the kernels yielded 2.08 per cent. of ash.

These analyses, the one of the ash, and the other of the seed and husk in their natural state, are sufficiently full for the purpose in view, and serve admirably to show the principal elements required in the growth of the Peanut plant. We see that albuminous matter and starch form a very large per cent., over three-fourths, of [Pg 13] the seed. Of course an article so rich in fat-forming ingredients, must be well suited for the food of man or beast. This explains why hogs fed on peanuts take on fat so readily. Nothing will change the appearance of a poor hog sooner than a diet of peanuts. The amount of oil in the seed—sixteen per cent., makes the Peanut one of the best oil-producing plants in the world.

Of the mineral constituents, potash forms by far the largest part— 44.73 per cent. Soda, magnesia, and phosphoric acid also enter quite largely into the composition of this plant. It will be noticed that common salt plays some part in the make-up of the Peanut.

Some may wonder at the small amount of lime reported to be present in the ash. This may be explained by stating that lime is not *per se* a manure, but a powerful chemical agent when applied to the soil, reducing inert matter into plant food. Lime appears to be the driving-wheel in the laboratory of the soil. Its presence is essential, but it does not do all the work itself. Of marl, the best fertilizer yet discovered for the Peanut, the principal ingredient of value, is carbonate of lime. Some of the Virginia marls range as high as seventy and eighty per cent. in carbonate of lime. This form of lime is very valuable for all agricultural purposes. Like its more caustic relative, it plays the part of a solvent and liberator, refines and vitalizes the soil, and causes other ingredients to perform their part in building up the framework of plants.

FOOTNOTE:

[1] While "goober" may be one of the names of the Peanut in some localities, the plant so-called in Georgia is *Amphicarpæa monoica*, a native leguminous plant with two kinds of flowers, one set always subterranean, and the other above ground. The under-ground flowers bear woody, rounded, one-seeded pods, with a seed closely resembling a bean. — Ed.

CHAPTER II.

PLANTING.

Soil, and Mode of Preparation.—A warm soil is required by the Peanut. A light, porous soil in which sand predominates, but not too sandy, warm and dry, and yet not too dry, but containing some moisture, and open to capillary circulation, suits the Peanut best. In all cases the soil most suitable for the Peanut must contain a certain amount of calcareous constituents. The color of the soil should be gray, with few or no traces of iron to stain the pods. As a rule, the brightest pods bring the most money, and as the color of the pods is always influenced by that of the soil in which they grow, it becomes a matter of importance to select that which is of the right description. Land of the above nature and color may be regarded as first-class for this crop. But let it be distinctly borne in mind, that unless it contains a goodly per-centage of lime in some form, in an available state, no land will produce paying crops of pods, although it may yield large and luxuriant vines. Of all the forms of lime, that supplied by the marls of the seaboard section appears to be the best.

But any soil that can be put into a friable condition, and kept so during the period of cultivation, will produce salable peanuts, provided it contains enough lime to insure solid pods. If it is known that a piece of land will produce sound corn, at the rate of from five to ten barrels per acre, the planter may rest satisfied, without further experiment, that it will yield from forty to [Pg 15] seventy-five or eighty bushels of peanuts. As the cultivation extends, and more land is needed for this crop, much of it is being put upon clayey soil, and when well cultivated, it generally produces heavy peanuts. Indeed, more pounds per acre may be grown upon some stiff lands than on any light soil, however calcareous. But clayey land, or such as is dark or tenacious, will impart a stain or dark color to the pods that is objectionable to buyers, and hence soils of this nature are generally avoided. A tenacious soil is also colder and more inert than a light one during the earlier part of the summer, and as the Peanut plant requires a rather long term of warm weather to insure full growth and maturity, a warmer and quicker soil is preferable. Buyers, however, are not now quite so particular as formerly in regard to color, and hence there is more inducement to plant on any ground that will yield good, solid peanuts, and it is being more frequently done.

But the actual or prospective peanut planter, who has an ash-colored or grayish soil, which is sandy and non-adhesive, is fortunate. If he will keep it well limed and trashed, or else rotate every fourth or fifth year with the Southern Field Pea, or other green crop, and marl, he will have land that will continue to produce paying crops of the brightest and most salable peanuts. There is an abundance of good peanut land all along the Atlantic seaboard, from New Jersey to Florida. Doubtless there is much of it in the Mississippi Valley, even as far north as the lake region, and on the Pacific coast from Oregon southward. There is no more reason for confining the cultivation of the Peanut to the narrow belts [Pg 16] at present occupied, than there is for limiting tobacco to the States of North Carolina and Virginia.

The quantity of lime or marl to use at one application depends very much on the nature of the soil and the amount of vegetable matter it contains. Generally, fifty bushels of lime, or one hundred and fifty bushels of marl is a safe application, but if the soil is quite thin, and contains but little vegetable mould, more than this at one time would be attended with risk. The safer plan is, to make several small annual applications of both marl, and vegetable matter, continuing this until a hundred and fifty bushels of lime, or two hundred and fifty, or three hundred bushels of marl have been applied. After this, no more calcareous matter will be needed in fifteen or twenty years. Land will bear large quantities of marl with perfect safety, if kept well stocked with some vegetable matter to subdue its caustic effects. But as most of the best peanut soil is deficient in this respect, the planter should begin cautiously, using small quantities until he has deepened his soil and supplied it with vegetable mould by trashing the land or turning in green crops.

In choosing land for a peanut crop, some attention should be paid to the previous crop. The Peanut requires a clean soil, one clear of roots, brush, stones, or rubbish of any kind, and hence it should follow some hoed crop, such as corn, cotton, or tobacco. In Virginia, corn land is generally preferred, and, as in the tide-water section, much of this land has been heavily marled, it commonly produces well.

The preparation of the soil for the Peanut is the same [Pg 17] as for corn, or any similar crop, except that more pains should be, and generally are taken, to get it in fine and mellow tilth. If it breaks up rough and turfy, as much land previously in corn is apt to do, it should be harrowed or dragged until it is fine. Generally, Virginia planters do not plow quite so deep for peanuts as they do for corn. This practice the writer believes to be unsound. Land should be plowed deep at the outset for all crops, whatever their nature or manner of growth. Deep plowing is a corrective of dry weather, and as drouth sometimes tells heavily on the Peanut plant, as was the case in the season of 1883, it is always well to plow deep, and give the moisture of the subsoil a chance to rise upward, and reach the roots during a dry spell. The formation of a fine, mellow seed bed, is all the preparation a peanut soil requires, previous to planting time, apart from the application of manures, which is spoken of elsewhere.

The Seed.—With the peanut crop, more than with almost any other, good seed is a matter of paramount importance. The seed sometimes fails to germinate well; before this fact can be discovered, and the ground re-seeded, unless the first planting was made quite early, the best season for planting will have passed, and the crop planted late will never be so good as it might have been. On the other hand, a very early planting doubles the risk of failure, in fact almost challenges failure by committing the seed to a soil too cold for germination and a quick growth. It is highly important, then, to have good seed, and to wait until both weather and soil are favorable for speedy germination and growth.

[Pg 18] In order to determine whether the seed will germinate well or not, let the planter begin to test them early in the spring. Let him take a dozen or two kernels that appear to be in quality a fair average of the whole lot of seed on hand, place them in a tumbler with some dampened cotton, or a piece of sponge, and set the tumbler in a warm place, where the heat is uniform, and high enough to start the germ in a few days. In a day or two, if the seeds are good, they will begin to swell, and the embryo plant will soon begin to grow. Thus, according to the number of seeds that have germinated out of the number tested, the planter can calculate the probable percentage of good seed. A glass of peanuts growing thus in dampened cotton, presents an interesting study, and is a pretty ornament for the sitting room.

But the planter must not rest satisfied with one trial. As soon as the out-of-door temperature will admit of it, he should try quite a number of the seeds in the open ground. Selecting a warm, sunny spot, he should plant from fifty to one hundred kernels, and shelter the place as much as possible from the cold winds. If these germinate well, the seed may be relied upon as good, and no further trial need be made. It is in this way that the Virginia planter tests his seed every season. About the first of April there is a great testing of the seed peanuts, and, although nearly every planter endeavors to save his own seed, the quantity of doubtful seed is generally great enough to cause a brisk demand for good seed at advanced prices. The method of saving seed peanuts will be given in a subsequent chapter.

Some weeks before planting time, the Virginia farmer, [Pg 19] who plants from fifty to a hundred bushels of peanuts, starts about having them shelled and assorted, preparatory to planting. This must be done with care, and females are mostly employed to perform this work. The pods are popped open with the fingers and thumb, care being taken not to split or bruise the kernel; all shrivelled and dark colored kernels are rejected. After they are shelled, the seed must be put into bags or baskets, a small quantity in each parcel, and set where there is a free circulation of air, until wanted for planting. If a large quantity is bulked together after being shelled, or if put in a close box or barrel, even in small quantities, they are liable to heat, and be prevented from germinating. This fact is the result of some costly experience on the part of many planters. Thus it becomes necessary to handle the seed with great care and circumspection throughout. From a bushel to a bushel and a half of peanuts in the hull, or pod, is estimated to be enough to plant one acre of ground, the quantity depending on the quality of the seed and the distance apart they are to be planted.

Time of Planting.—In Virginia, the first twenty days in May is regarded as, in the main, the most suitable time for planting. Some plant as early as the last week in April, and the seasons frequently favor this early start, and the crop does well. More, however, plant in June than in April, and sometimes planting is delayed until the middle or last of June. On warm and dry land, there is no great risk in planting the first week in May, but on colder land, the planter should wait until the ground has been warmed by the sun, say the latter part [Pg 20] of the same month. If the farmer has reason to hope for a week or ten days of mild, fair weather, he may risk a planting quite early, as in that time the seed ought to germinate, and come up sufficiently to make it sure that it will grow. Once up, the plant will hold its own, and though cold rains or winds may retard its growth, and cause it to turn yellow, it will start anew with the first spell of sunny weather, and rapidly change color to its normal green. The above dates apply to the latitude of Virginia. In the far south, peanut planting begins early in April, while north of Virginia, the first half of June would, in most seasons, be quite early enough to commit the seed to the earth. It should not be done anywhere until all danger from frost is passed for the season. A very slight frost will destroy the Peanut.

How to Plant.—I come now to consider the mode of planting. Here no very inflexible rules can be given. Practice varies greatly, almost every planter differing more or less from his brother planters. The chief points are, to get the seed into the ground at suitable distances apart both ways, to have the seed, after it is planted, raised slightly above the general level, and to have the soil so free from clods that there will be nothing to hinder the young plant from pushing through after it has started. Any mode of planting that will secure these ends will effect the purpose.

If the ground has been once plowed in the early spring, let it be plowed again only a few days before planting time, and if at all rough, or cloddy, have it harrowed until in fine tilth. When ready to plant, draw furrows [Pg 21] the same as for corn, two and a half or three feet apart. If the land is fresh and strong, and never before in peanuts, make the rows at least three feet apart. After a year or two on the same ground, peanut vines will not grow so large as at first, and need not be so far apart, either from row to row, or from hill to hill. When the land is thin, some plant as near as twenty-seven inches from row to row, and twelve inches from hill to hill.

If any fertilizer is to be used, let it be put in the furrow before the ridge is formed; a man or boy following the plow and spreading the fertilizer by hand. A small ridge is then formed by lapping two furrows over the drill with the turn plow, after which the knocker and dotter follow, one leveling the ridge, and the other dotting the row by making little depressions in the soil the proper distance apart for the seeds.

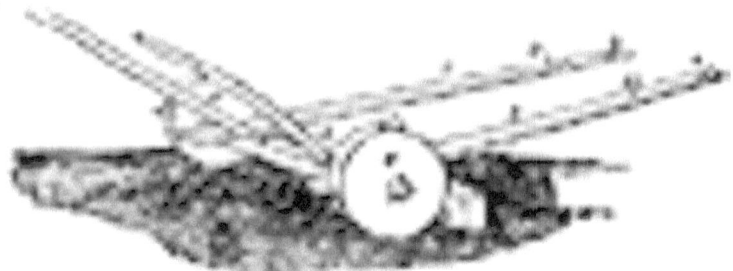

Fig. 2.—THE KNOCKER AND DOTTER COMBINED.

The Knocker and Dotter.—Sometimes the knocker and dotter are combined in one, and it is withal a unique implement. Always home-made, it partakes of all the native roughness and varied ingenuity of the Southern planter. The engraving, figure 2, will illustrate the mode of constructing this implement. Two pieces of timber are sawed from a log to serve as wheels, such wood being selected as does not split easily. The diameter of the [Pg 22] wheel is made the same as the desired distance between the hills, and three wooden pins are inserted equi-distant in the circumference, so that the wheels will make three dots, or signs, for planting, at each revolution. These wheels are connected by an axle, and set the same distance apart the rows are to be asunder. Two shafts are pinned to the axle, and braced in front of the wheels to keep them steady. A piece of heavy scantling, or a log of wood, six inches in diameter, is secured to the under side of the shafts just in front of the wheels. This is the knocker, and serves to level the ridge before the wheels. Properly adjusted, it does beautiful work, and leaves a flat, smooth ridge, in fine condition for the seed. The wheels pass along on the leveled ridge, making the dots, as shown in figure 2. Handles are fixed to the implement to enable the plowman to keep it in proper place, and for convenience in turning. One horse is fastened to this implement, and two rows are prepared for planting at the same time. This utensil would be troublesome to use in an orchard, or on stumpy ground. Peanuts, however, should always be planted on open ground clear of all impediments. Instead of the knocker and dotter combined, many planters omit the wheels, and make a separate implement with one wheel and a handle, to work by hand, as represented in figure 3. [Pg 23] This can be run among trees and stumps. It resembles a wheelbarrow without the body.

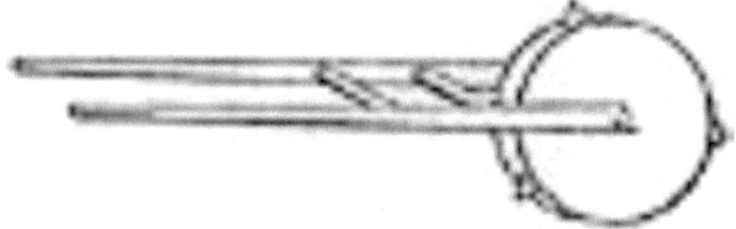

Fig. 3.—THE DOTTER.

Hands—women, children, or men, follow the dotter, dropping a seed in each mark or depression, and carefully covering it with the foot, by pressing enough soil into the hole to just fill it. The holes are made one and a half to two inches deep, and the hands are cautioned not to get the seed covered deeper than that. One inch is deep enough to plant, if the soil is moist, but if quite dry the seed may be put deeper. Proceeding in this way, covering first with one foot and then with the other, the planters get on quite rapidly, although the hills are so near together. The planting is not at all tedious after one gets the knack of it, and is light and pleasant work. Some planters put two kernels instead of one in each hill, to insure a stand, but this practice increases the cost considerably, and is by no means general. After the seeds are planted they are very slightly, if at all, above the common level. In a week or ten days from the time of planting, the seeds will begin to heave or crack the ground, which shows that the germ has started, and greatly relieves the anxiety of the planter. Then, by counting the number of signs in a hundred hills, the farmer readily calculates what kind of a stand he will probably have.

Fertilizers.—We have already intimated that a calcareous soil is indispensable to successful Peanut culture. If the soil is not calcareous by nature, it must be made so artificially. Hence the proper fertilizer to use is one that contains a large per cent. of lime in some of its [Pg 24] forms, as the carbonate, the phosphate, the nitrate, or the sulphate, or the chloride of calcium. Recently, the sulphate of lime (gypsum), has been employed, even on limed or marled land, and its use has been attended with good results. Animal and nitrogenous manures are not suited to the crop. Such fertilizers produce a heavy growth of vines, but there will be no full, solid pods unless lime in some form is also present. Marl has been found to be the one specific fertilizer for the Peanut plant—better than any other form of lime; and the chief element of value in marl has been shown to be the carbonate of lime. Some Virginia marls contain as high as seventy-five or eighty per cent. of the carbonate, and all of them range over twenty-five or thirty per cent. Now, marl is plentiful and cheap all along the Atlantic seaboard, from New Jersey to Florida, the beds lying side by side of, and intersecting, the very land that is the best adapted to the Peanut—a rare and fortunate coincidence, that planters are learning to fully appreciate. And were it not that the New Jersey land-owner finds it more profitable to raise fruits and vegetables for the two great cities that lie on either hand of him, even he would find the Peanut to be a paying crop. With his warm, light sand and green marl, he could easily raise them. I mention this as one of the possibilities of the Peanut, though not likely to be realized for the reason named.

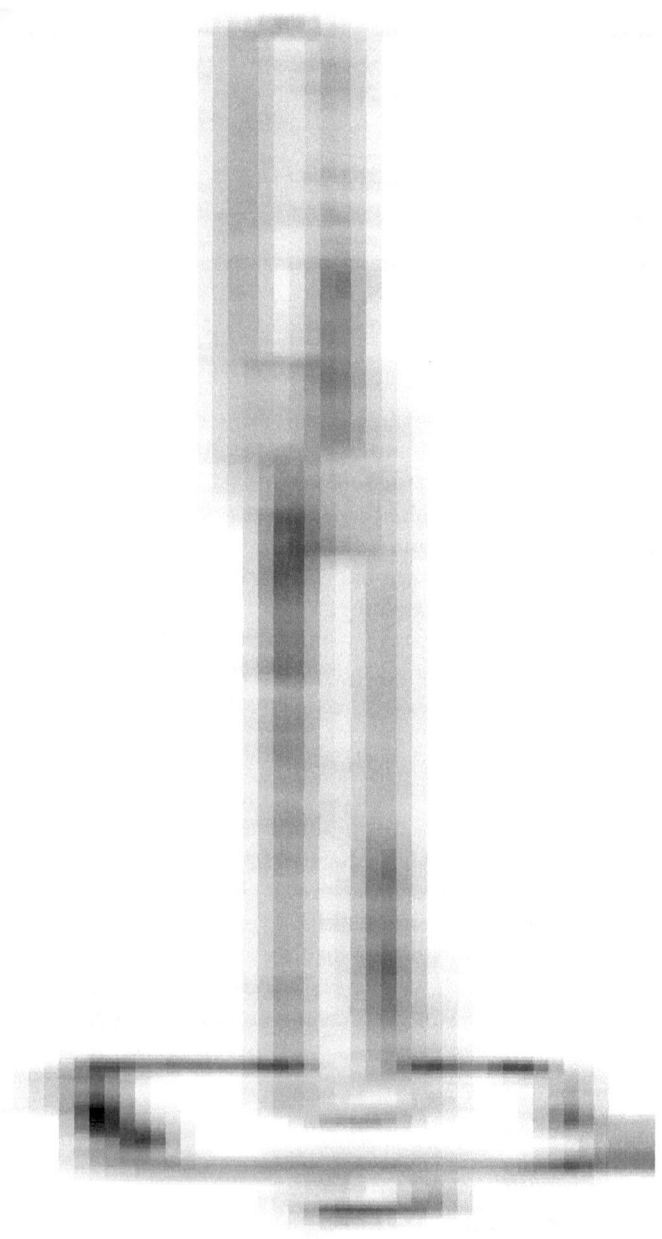

Fig. 4.—STAKE.

Replanting.—In about two weeks from planting, if the weather has been mild, the young plants should be large enough to show where replanting is necessary. The planter goes along the row, making slight [Pg 25] depressions with his heel at all the missing hills, drops a pea therein, and covers it with the foot, the same way as at the first. Instead of making depressions with the heel, some use a long stake, an inch or two in diameter, to the lower end of which is affixed a piece of plank, fastened two inches from the end, and four or five inches long (fig. 4). This is used for punching the holes, and the piece of plank near the end prevents it from making the impression too deep. This is another of the inventions of the Virginia Peanut-planter; so true is it that "necessity is the mother of invention," a new crop calls for new devices for its successful and profitable cultivation.

In replanting, it is well to put two or more kernels to the hill, as the season will be getting late, and no time should be lost in securing a good stand. There can be no subsequent replanting with any profit.

Moles and other Depredators.—The Peanut-planter has to contend with many enemies. In many cases moles are exceedingly destructive to the planted seed, burrowing along the rows, and eating the seed, hill by hill. Often raccoons, foxes, and squirrels grabble them up. And everywhere the larger birds, such as crows, doves, and partridges come in for a share of the seed, and annoy and hinder the farmer very much. There is no remedy but ceaseless vigilance. The planter must go armed at every turn to protect his crop. Sometimes planters tar the seed to prevent the moles, etc., from [Pg 26] destroying them. It perhaps has some tendency to check the depredations, but does not prevent them entirely. Coal tar is oftenest used for the purpose, a half pint being enough to smear a bushel of seed. The seeds are afterwards rolled in dry earth to prevent adhesion and trouble in planting. Traps, guns, and scarecrows are resorted to with varying success, but if the depredators are numerous, the planter is generally the vanquished party.

The Critical Period.—The first four or five weeks after the planting of this crop is its most critical period, and nothing but a good stand and the approach of warm weather will relieve the planter of his anxiety. At the first, many fears are reasonably entertained that the seed will not germinate well. And even should a pretty fair percentage of the seed come up, cold and rainy weather may still seriously retard the growth of the plants, or the numerous depredators that have been named may so far reduce the number of hills as to greatly curtail the yield per acre. The very young Peanut is among the tenderest of plants, and a very slight mishap will serve to destroy or permanently injure it. Several days of cold weather at this period will make the struggling plants look pale and sickly, and if warm suns are too long delayed, many plants will fail altogether.

Backward springs are a great drawback in the cultivation of this crop, and cause many farmers to delay planting until it is certain warm weather cannot be many days off. If the planter could always be sure of his seed, this would be the better plan, but if these late plantings fail [Pg 27] to come up well, the season is too far advanced for replanted seed to make a crop. Further north than Virginia, however, it would, we think, be decidedly better to put off planting until both soil and air are warm enough to insure quick germination, and then, instead of replanting the missing hills with Peanuts, plant beans or field peas instead. If the planter can get through the first month successfully, he lays aside his fears, and enters upon his work with renewed hope and energy. To a recital of this work—the work of cultivation, we now invite the reader's attention.

CHAPTER III.

CULTIVATION.

First Plowing and Weeding.—Usually, the cultivation of the Peanut begins by first siding the rows with a turn-plow, small mould-board attached, by which the soil is thrown from the plants, and lapped into a small ridge in the middle of the balk. Care is taken to run the plow quite near to the plants, so as to leave as little as possible for the hoe to do. The hoes follow the plow, removing the grass between the hills, if any, and loosening the soil about the plants. Sometimes, however, in case the plants begin to get quite grassy very early in the season, the sides of the ridges are first scraped off with the hoe, the operator moving backward, and clearing off one side at a time. This removes the grass pretty well, but does not loosen the soil about the plants. If this [Pg 28] method is pursued, the plow should be put on in a week from that time, to break the hard crust that will have been formed, and to let in the air and heat to the roots of the plants.

If the first plan is followed, the missing hills may be replanted, if the former replanting has had time to come up, but otherwise the ground about the missing hills should not be disturbed. This, however, should depend upon the time at which the weeding begins. If very late, it is useless to replant.

The time for the first weeding must depend somewhat on the nature of the soil and the quantity of grass that may have sprung up since planting. Usually the first working should begin by the time the plants are two weeks old, but if the land is mellow and there is but little grass, the work may be put off a week longer. But if rains have occurred and a crust has formed, and especially if grass is coming on rapidly, the planter should not wait for the plants to attain a certain age and size, but should proceed to work the crop as soon as the plants are clearly out of the ground, and have put forth one or two branches. Any practical farmer who knows how to plow and weed young corn, will not be likely to err very far in working a crop of peanuts. The operation is simple enough, the two points being to clear away the grass and make the soil fine and loose around the plants. Any plan of working that will secure these ends, will accomplish the purpose.

Subsequent Workings.—The second plowing may be done with a cultivator, running twice in the row. This [Pg 29] will level the ridge in the middle of the balk, make the soil loose and fine, and bring the loose earth up close to the plants, which will make easy and nice work for the hands with the hoes unless there is a great deal of grass. The second plowing and weeding is the most important working the crop receives, and it is highly important that it be done well. By this time (last of June), the days are long and hot, the grass everywhere is growing apace, and the Peanut must be kept growing too. The plants have now attained a size ranging from that of a saucer to that of a breakfast plate, and there will be some hand-picking of grass necessary, because some of it will be found growing too near the plants to be cut away with the hoe. If there is very little grass, the work goes on smoothly enough, the hoes proceed quite rapidly, three hands keeping up with one plow, and finishing about two acres a day.

The third plowing may be given with a shovel or cotton-plow, or with the cultivator, again running twice in the row. The hoes need not follow at this plowing, but may wait until the fourth plowing, done usually toward the middle or last of July, or about the time the vines are a foot in diameter, and are sending down their peduncles, or stems, on which the young pods are forming. The plants begin to blossom by the first of July or before, and continue to flower for more than a month. The pods begin to form very soon after the flower appears, and by the time of the last weeding great care must be taken not to cut the stems. For this reason the hoes cannot proceed as fast as at the last weeding, and if there is much grass growing up through the vines to be hand [Pg 30] picked, this working is tedious and laborious enough, and tires to the utmost the patience and endurance of the laborer. In fact, this is the worst period in the cultivation of the peanut crop. The weather is hot, close, and enervating; the frequent stooping and picking makes it doubly laborious; and, on account of the size the vines have attained, the plow must necessarily leave a wider surface for the hoe to go over. All this makes greatly against the hoe hands.

It is no wonder, then, that, with laborers, many of whom are disposed to shirk their duty, the last working is too often poorly and inefficiently done. With more reliable labor, such as is to be had in the Northern and border States, better success would be easily attainable.

The third weeding is the last working with the hoe that the crop receives, and next to the last usually given it with the plow. The Virginia planter, as a rule, stops weeding by the first of August, or as soon as the vines have well met along the row, and have sent down a goodly number of young pods. If there is any subsequent removal of grass, it is done by picking it out by hand, in order not to interfere with the pod stems. But after the last weeding, say in a week or ten days, one more plowing is usually given, generally with the cultivator or shovel-plow, run once in the row. This throws the soil up under the extremities of the vines, leaving the row of plants on a nice flat bed and a water furrow in the middle of the balk.

The reader will observe that the cultivation required for the Peanut is such as will keep the soil mellow and loose on the surface and clear of grass, especially about [Pg 31] the vines or plants. Any method of weeding and plowing that will secure these ends, will serve the purpose. Accordingly, there is a considerable diversity of practice in this particular, both as to the mode of plowing, times of working the crop, and implements used. The cultivation, however, is as easy and simple as that commonly bestowed on Indian corn or beans, but must be a little more thorough and painstaking. That is all. None need shrink from planting this crop through any apprehension that they will not work it properly. The three essential points are: keep the soil loose, the grass down, and do no harm to the young pods as they are forming on the vine.

Implements.—This topic has been, in a measure, anticipated, allusion having already been made to the implements to be used in the cultivation of this crop. A few additional remarks, however, may not be out of place.

The weeders should be armed with the best steel hoes, with factory-made helves of ash, light and slightly flexible. The superiority of this hoe—usually called the "goose-neck hoe" in Virginia—over the old style of weeding hoe, with the heavy and stiff home-made helve, cannot be estimated, except by those who have tried both. The same hand can perform an eighth more labor in a day with the light steel hoe, and do it better, and with more ease to himself. The "goose-neck" will last two or three seasons, costs but little more than the other kind, comes ready for work, and is, therefore, very cheap. The blades should be kept sharp by repeated filing.

[Pg 32] With us the first plowing is generally done with the turn-plow, with a small mould-board attached, throwing the earth into the balk. For the second plowing, the cultivator or cotton-plow, is used, either one of which does fine work on smooth land, and makes it quite easy for the hoe hands. The third plowing is commonly performed with the cultivator, but if the ground is rough, the turn-plow will answer better. It is not common, however, to plant peanuts on very rough ground. For the fourth and fifth plowings the cultivator or shovel-plow is used. But should the crop get very grassy, (which should never be permitted), the turn-plow, with large mould-board attached, is used, in order to cover up as much of the grass as possible. This makes a large and objectionable ridge in the balk, but it is the best way to conquer the grass when it gets too strong a hold. The hoes follow the plow, and scrape off the remaining grass, except that near the plants, into the balk. Bunches of grass that have grown up among the vines have to be pulled out by hand. Thus, it will be seen that there is no plow made especially for cultivating the peanut crop, the same plows and implements that are used for other and general farming purposes answering equally well for the cultivation of this crop also.

When Cultivation should Cease.—When the peanut vines have interlocked considerably along the rows, and have almost, or quite met across the balks, it is high time to cease cultivating them. When the vines are large, the cultivator or plow will tear and bruise them more or less, sometimes breaking off large branches, and, [Pg 33] of course, destroying a number of pods. If there is not room for the plow to pass without pulling out the young peanuts and harming the vines, it should be taken off the field and the crop left to take care of itself. So long as the vines remain small, the crop may be worked to some extent, provided always that care be taken not to molest the stems that have penetrated the soil. Every one of these that is harmed now is a peanut lost. In Virginia, two months—June and July—covers the period of cultivation for the peanut crop, and it cannot be extended much beyond this time without some risk. In fact, a crop that has been faithfully worked during this time will not require anything more, and any extra labor is as good as thrown away.

Insect Enemies.—Fortunately for the planter of peanuts, there is scarcely an insect that does them any material harm. At least, such has been the case, so far, in Virginia. What subsequent years may bring, is, of course, unknown. But up to the present, no insect has ever caused any extensive injury to this crop. It is true that ants do sometimes destroy a few hills on certain soils, by sucking the cotyledons of the plant before it has attained any considerable size and strength. But this is, by no means, general. Even the voracious and ubiquitous Colorado Beetle manifests no taste for this plant, although it has had abundant opportunity to test its edible qualities. To the credit of insects generally, be it said, they are not omnivorous.

Effects of Cold.—The effect of severe and prolonged cold on the Peanut plant in the early part of the season, [Pg 34] is often quite manifest. Cool nights and cold rains are much dreaded, they cause the plants to turn yellow and look sickly. The vines make little or no growth, the leaves become spotted and curled, as if they had been touched by fire, and the whole plant gets into that unthrifty looking state denominated, in the local parlance of the planter, "the pouts." But let a few days of warm sun occur, and all is speedily changed. The plants assume a fresh and lively green, and their growth is now rapid until they reach maturity.

Effects of Drouth.—A very dry spring would cause the Peanut to come up badly, and would, therefore, seriously affect the crop. Such an occurrence, however, is very rare in Virginia, as well as in the country generally, and is not regarded with much apprehension. If the plant is once well established in the soil, being tap-rooted, it can stand a good deal of dry weather. It takes a long period of extremely dry weather to materially injure this crop. Such a season did occur in 1883, and the consequence was a great many blasted pods and a short crop. Generally, moderately dry summers are looked upon with favor by the planter, inasmuch as seasons of this kind enable him to keep the crop clean of grass at much less cost. Just here we would repeat what we said in Chapter II, in relation to deep plowing preparatory to planting. With a soil deeply broken in the outset, the Peanut will withstand successfully any period of dry weather ever likely to occur in this country. It has been noticed that the crops that suffer the most from drouths are those planted on land not well prepared, or [Pg 35] in orchards of growing trees, which necessarily extract a great deal of moisture from the soil. Even in a season as severe as that of 1883, peanuts planted on a deep, mellow soil out of the reach of trees, did well, and were well seeded and filled. Deep preparation of the soil, then, is a corrective of drouth for this crop, as well as for any other. With this simple precaution, no great apprehension need be entertained of the effects of dry weather. Let the planter but do his part in preparation and cultivation, and nature will be sure to respond with liberal, if not overflowing crops. The corn-planter has more to fear from dry weather than the peanut-planter.

Appearance at this Period.—The appearance of a thrifty crop of peanuts at the time of maturity, or a little after the last weeding, is simply magnificent. The vines have now met in both directions, and the whole field, from a little distance, looks as if covered with a carpet of velvet-plush. Nothing obstructs the view. The vines lie close on the soil, and the eye reaches every nook and corner of the field, and takes in the whole panorama at one glance. Few other crops afford so clear or so pleasing a prospect. Indian corn, in the tender green of summer, is a beautiful object to look upon, but it shuts out all view of distant parts of the farm. The golden wheat, as it bends to the passing breeze, is also beautiful, but one must go around it and not through it. A field of cotton, as the open bolls display the snowy lint, is a sight to please the admirer of nature, but it lacks the setting of green that is always pleasing [Pg 36] to the eye. The peanut crop surpasses them all in beauty. It presents an air of freedom, of repose, of life, and of security from harm, of which no other can boast.

Such is the crop to which we have invited the reader's attention, and the planting and cultivation of which we have endeavored to describe. Having proceeded thus far, let us pause a moment, as the writer has done, time and again, to survey the beautiful prospect of a field of peanuts in full maturity. There it is, a literal carpet of living green, covering acres on acres of mother earth, and beneath its velvet folds is quietly growing the wealth that is to make its owner independent, and by means of which the planter's family is to secure most of the necessaries and comforts of life. No crop outside of the market gardens, yields so much actual cash per acre as this. No wonder, then, that it readily becomes popular with all who try it, and that it never loses ground wherever introduced under favorable circumstances.

An interval of about two months now elapses, during which the crop requires no attention. The seed pods are filling and maturing, and the whole plant is ripening for the harvest.

[Pg 37]

CHAPTER IV.

HARVESTING.

When to begin Harvesting.—We come now to the laborious and often difficult work of harvesting the peanut crop. We say difficult, for often rainy or other unpropitious weather at this period, makes it exceedingly hard to save the crop in good condition, and prevent the pods from becoming dark or spotted. Ordinarily, the harvesting should not begin so long as mild and growing weather continues, even though October may be far spent. It is important, of course, to get as many firm, matured pods on a vine as possible, and the longer the weather holds favorable for this, the more pods, as a rule, will there be.

If, however, the crop has been planted early, and the leaves begin to fall from the vines, it is better to start the plow and dig the crop at once. When the Peanut plant gets fully matured, it is very apt to begin to cast its leaves, especially on ground that has been planted in peanuts often before. After the leaves fall off, the vines are of very little value as hay, and as most planters consider them excellent provender, they make it a point to harvest the crop in time to secure good hay. For the same reason, effort is made to dig and shock the vines before a killing frost occurs. Frost spoils the vines for fodder, though it does no harm to the pods, unless it be for seed. Some suppose that seed taken from frost-bitten vines will not come up well.

In the latitude of Virginia the usual time for digging [Pg 38] the peanut crop is the second and third weeks in October. That is, the great bulk of the crop is dug about this time, though some start the first week in that month, and others wait until the close, unless driven to start earlier by the weather. In rare cases, some planters dig by the twenty-fifth of September, but it is generally believed that all who start thus early lose more in weight and yield than they gain in time or price. Six or ten days of mild weather at this stage of the crop, will make an appreciable difference in the yield, and if the peanuts can remain in the ground until the latter part of October, there will be very few saps, or immature pods. But, in whatever latitude the planter may reside, the general rule should be, to dig before a killing frost occurs.

Mode of Harvesting.—In Virginia, the general practice is as follows: First, plow the peanuts with a point having a long, narrow wing, and a small mould-board, so that the vines will be loosened without having any earth thrown upon them. The plow passes along on both sides of the rows, just near enough for the wing to fairly reach the tap-root, which it severs. Care is taken to put the plow deep enough to pass under the pods without severing them from the vines. This is important, as most of the detached pods are lost, and if the work is slovenly done, the loss will be great.

Hands with pitchforks follow the plow, lift the vines from the loose soil, shake them well to get the earth off, and then lay them down, either singly or in small piles, to remain a day or two to wilt and cure in the sun. This is light work, and can be done rapidly, two hands being [Pg 39] enough to keep up with one plow. If rain is feared, it is best to lay the vines down singly after shaking them, for, when in piles, if rain occurs, and the weather is warm, the pods are apt to speck and mildew before the vines can dry out. A rain falling on the pods after they are dug, and before they are shocked, does no harm, if the sun comes out soon to dry them before they can mildew.

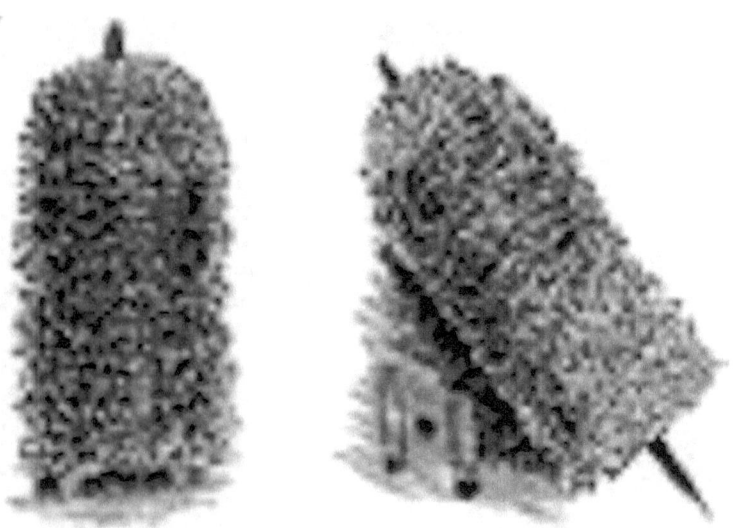

Fig. 5.—SHOCK STANDING.　　Fig. 6.—SHOCK REMOVED.

Instead of leaving the vines on the ground a day or two to cure, many shock them up at once. If the vines are perfectly dry, this is as good a plan as any. But if the weather should be warm, and the vines are wet with dew or rain when put up, they will be sure to heat, and the pods will turn dark. In cold weather the vines may be shocked both green and wet without risk.

The method of shocking the Peanuts will be understood from figure 5, which represents a shock as it [Pg 40] stands in the field. A shock as it is taken down for picking is shown in figure 6. The vines are first laid together in piles, about as much as one can handily carry on the fork at one time, three rows being put in one. The stakes, which have been previously prepared, are then set in the ground proper distances apart, and two billets of wood, four or five inches in diameter and two feet long, are placed beside each stake to keep the vines off the ground. A handful of vines is then laid, pods up, on one side of the stake for a bed, and the same on the other side. After this the vines are put on, pods down. The first are inverted to keep the pods off the ground, though this is a matter of trifling importance, if the billets of wood are large enough. The successive handfuls of vines are laid up with care, keeping the shock level, lapping the vines, and placing them on every side to make the work even. As the work progresses the vines may be pressed down with the hands, and the shocks are finished off round at top, the better to shed the water. No cap or covering for the shocks is used, though much would frequently be saved, could a cheap one be had. A board nailed on the top of the stakes would protect the top layer very much, and yet the planter who should adopt it would doubtless be laughed at.

A fast hand can put up fifty or sixty shocks a day, with a boy to bring up the vines and assist in planting the stakes. Some shockers use the fork to lay up the vines, especially toward the top. The shocks are put up one in a place wherever needed, so as to make the work convenient for the carrier. Some, however, put three or more shocks together, as suits their fancy, in which case [Pg 41] fence rails are usually employed to build the shocks upon.

The above method is generally practised, but there are many variations in almost every detail. We have endeavored to give a clear idea of a safe method.

Why Cured in the Field.—Perhaps some reader unacquainted with the cultivation of the Peanut, may ask: Why all this trouble to shock and cure the crop in the field? Why not pick the pods from the vines as soon as they are dug, and cure the peanuts on scaffolds, or elsewhere, and cure the vines on the ground, like hay?

We answer, because the pods cure better in the shock than in any other way. They get dry sooner, and make heavier and brighter peanuts than could possibly be the case, were they gathered at once, and spread, even in very thin layers, on scaffolds to dry. Besides, as rain on the pods when they are about half cured, or during the process of curing, would be very harmful, it is found best to protect the pods by covering them in shock. They can get more air in shock than if spread on a scaffold, and a free circulation of air about them is important. A scaffold close enough to hold the pods would exclude the air in every direction, except from above. When shocks are put up well, the pods are very effectually protected, except a few on the top, and in about ten days are cured nice and bright, and ready to be picked off. The shocks may remain in the field many weeks, subject to repeated rains, without material injury. Of course rains of several days continuance would damage the peanuts more or less. It is best therefore, on this account, and because [Pg 42] of the numerous depredators that prey upon the crop while it remains in the field, to house it as soon as sufficiently cured to render it certain the pods will not heat and spoil when in bulk.

Depredators.—The creatures of the animal kingdom that levy their tax on the unwilling planter, and come in for a share—and often a large share—of the peanut crop, are of many kinds, and numerous in all. Of quadrupeds, the deer, fox, raccoon, squirrel, and sometimes even the dog, are more or less destructive; the raccoon, squirrel, and fox are particularly so, beginning their inroads early in the fall by scratching up the immature pods, and continuing their thefts daily and nightly as long as any remain in the field. In some localities, these animals are exceedingly annoying, and occasion great loss unless their depredations can be checked.

Next to the animals named, birds are most destructive, while the peanuts are in shock. Such birds as the blue-jay, crow, partridge, yellow hammer, wild turkey, and blackbird, coming, as some of them do, not singly, but in companies and flocks of hundreds and thousands at a time, carry off vast quantities, unless the planter is always on the alert, gun in hand, ready to meet them at every turn. Near the James, and other large rivers, it is a common occurrence to see, not thousands only, but tens of thousands of blackbirds in a single field at one time. They often go in flocks covering acres on acres of ground, and with their ceaseless activity and endless trilling, present an appearance of which city-bred people can form no adequate idea. Of course they destroy a [Pg 43] vast amount of peanuts in a short time, unless speedily driven off.

There are also several species of field rats and mice, together with the domestic rats and mice that get into the shocks to feed on the pods, where they remain until disturbed by the pickers. Everything seems fond of the Peanut after it is made, and if the planter escapes the insect enemies in the summer, the exemption is more than offset by the numerous and voracious depredators of the fall and winter.

And against most of them, there is no effective remedy, the planter cannot watch his crop all the time, and traps are hardly worth using. It is true, something may be done with steel traps for such animals as the fox, raccoon, and squirrel. But for the rest, despatch in removing the crop from the field, is the only certain preventive. Even then the planter does not entirely escape, for rats and mice follow him within doors, and riot in luxurious living so long as a single shock remains undisturbed. Perhaps no crop the Southern farmer grows is subject to heavier or oftener repeated losses than the Peanut. Yet, despite it all, it is a crop that often pays very handsome returns. It has been, and is, the sheet anchor of many an East Virginia farmer, and if prices hold up, will continue to be, so long as there are lands here that will produce thirty bushels of peanuts to the acre. This is but the minimum; the maximum is not known; a hundred and thirty bushels per acre has been attained.

Detached Peanuts.—In the process of digging and shocking peanuts, many pods must necessarily become [Pg 44] detached from the vines. Some of these remain in the soil, out of sight, and numbers more are scattered over the ground, from one side of the field to the other. If the vines are fully matured, and have changed color or shed their leaves, and especially if frost has touched them, the pods come off much more freely than if the vines are still green, or scarcely done growing. Generally, the detached pods are the best of the crop, being those first matured, and which are therefore solid and heavy.

Of course these peanuts must not be lost. Women and children are employed to pick them up at so much per bushel. If it is found that many pods remain in the ground, a cultivator or light plow is run along the rows to bring them in sight. In this way the most of the loose peanuts are saved. Still, numbers will be left in the ground. The planter is at no loss, however, to secure these also, which he does by turning his fattening hogs on the ground as soon as he can remove the crop from the field. Hogs are exceedingly fond of the Peanut, and as soon as they find them out, they will continue to root for them as long as one can be had. Frequently, every square yard of large fields, will be burrowed over by the hogs in their search for the detached peanuts. No crop the planter grows will fatten a hog so quickly as the Peanut. Thus in the harvesting of this beautiful and profitable crop, nothing is allowed to be lost.

Saving Seed Peanuts.—It now remains to say something of the method of saving seed peanuts. Every step in this process must have in view one principal point—keeping the pods from becoming the least heated, either [Pg 45] in shock or in bulk. Perfect and continued ventilation must be secured. The vines should not be shocked while green, nor the pods kept in large bulk after being picked off. Neither should the vines be touched by frost, either before or after being dug.

It is customary to dig and shake the vines as usual, and leave them in the field four or five days, or a week, before they are either piled or shocked. In this time, if the weather is fair, the vines will be so nearly cured that not enough moisture will remain in them to create a heat, even in very warm weather, and they may then be shocked with perfect safety, after which they should remain in the field until thoroughly dry. Rain falling on the vines while they are lying in the field, does no harm, except it be to turn the pods a little dark, which circumstance makes no difference with seed peanuts.

When the seeds are picked off, keep them in baskets until ready to spread them in a cool, dry room, where they will be exposed to a free circulation of air. In no case should they be in bulk. Spread them thinly in some loft, where the air will reach them, and where they will be secure from rats and mice. They may be stored in sacks the same as for sale, and laid in an airy room to remain all winter. They should not be kept in a room where there is a stove, or one subject to currents of hot air.

These suggestions embody all that need be done to secure good seed. If peanuts are fully cured when picked off, and are not kept too close, they will prove good seed, unless there is some radical defect of the germ or vital powers. Keep them from heating, and they will [Pg 46] germinate and grow as readily as corn. Every planter may, and should, save his own seed. According to the number of acres that he thinks of planting, let him provide two bushels of seed (or forty-four pounds in the hull), for each acre, and he will have enough and some to spare.

CHAPTER V.

MARKETING.

It requires as much judgment to market a crop well, as it does to raise and harvest it, and often more. Unfortunately, the majority of planters are sadly deficient in that knowledge of commercial life, which would make them masters of the situation. Too often they are bound by lien or mortgage, or else they have run up a heavy bill at the country store, and when the crop is made and ready for market, they are obliged to sell forthwith. Generally too, this is the very time when prices are lowest, and so the planter is obliged to part with the fruits of his labor at the most unfavorable rates, and allow the middlemen to pocket the profits. It is only by careful economy and prudent management, on the part of each planter for himself, that this evil is to be corrected. Without entering into the details of commercial affairs, we will endeavor to show the planter how he may go into market with his crop, prepared to command the best prices. To this end, it is essential that he have his crop in the best marketable condition, remembering that a good article always sells well.

[Pg 47] **Picking off the Peanuts.**—This part of the work, usually done by women and children, may make or spoil the sale of the entire crop. If stems are gathered with the pods, and good, bad, and indifferent are all lumped together, with leaves and trash thrown in for good measure, a great deal of assorting and cleaning will subsequently be required, or else the sale of the crop will be impaired to the extent of one or two cents to the pound. In picking, the stems should be rejected, and the saps and inferior pods, if gathered at all, be kept apart from the rest. Only the best, brightest, and soundest pods should go into the A, No. 1's, and these, if clean of earth and trash, will always bring top prices. The saps also will sell, at lower rates. It is the neglect of these few precautions that so sadly curtails the bill of sale of many a planter. If planters would offer pickers extra inducements for clean pods, this difficulty would, to a great extent, be obviated. When the same price is paid for all, without regard to the manner of picking, a premium is offered for slovenly work, and the careless get better paid than the painstaking.

In picking, the pops should be refused altogether, and the saps and very dark pods go by themselves. Many planters, however, leave the saps on the vines, saving the best only. The saps, however, will sell, either in pod or shelled, and if numerous, will more than pay for picking them. It is, therefore, so much gained. It must be confessed, however, that the presence of a good many saps on the vines, makes them much more valuable as feed.

Just here let us explain that "pops" are pods that [Pg 48] have attained full size and firmness, but which are minus the seed. Dry weather, and the lack of calcareous manures in the soil, will cause many pops. "Saps" are immature pods, the last to form on the vine, and which might become good peanuts if they could have a longer period of growing weather. The presence of pops in the marketable peanuts is very detrimental to their sale, and hence should be carefully rejected in picking. Saps also are detrimental, but to a less extent than pops.

Price paid Pickers.—The price paid pickers varies somewhat from one season to another, according to the quality of the peanuts, and the market price received for them. Hands commonly board themselves, and receive so much per bushel for picking. Of late years, the price has stood pretty uniformly, at twelve to fifteen cents per bushel. The peanuts are either measured or weighed. If weighed, twenty-four pounds are counted as a bushel in the first part of the season, the extra two pounds being taken to make up for the subsequent loss in weight. If a hand is boarded by the owner of the crop, he gets but ten cents a bushel for picking. A fast hand will pick from four to six bushels a day, the children are just as likely to do this as grown people. Hence, at this season of the year, women and children earn what is considered pretty fair wages. Under the most favorable circumstances, the best hands will pick seven bushels a day. Very much depends, however, on the quality of the peanuts, and something also on the weather. In very dry weather, the stems come off with the pod, and pickers cannot do as well.

[Pg 49] **Cleaning and Bagging.**—After the peanuts are picked off, they should be cleaned, before being sacked. The object of this, of course, is to rid them of the earth that may still be adhering to them. It makes the hull look cleaner, and brighter also, and thus enhances the sale. Formerly, the planter made his own cleaning machine, but recently, since the starting of what are called "Peanut factories," the planter very seldom runs his peanuts through any machine at all, but sells them just as they are picked. Being thus rid of much trouble and labor, it is doubtful whether it would now pay the planter to clean his peanuts, as he once did. The price paid for them now, is almost as much as he would realize, were he to take ever so much pains in cleaning them.

Fig. 7.—VIRGINIA PEANUT CLEANING MACHINE.

But as the reader in other parts of the country, may desire to know something of the mode of cleaning peanuts at home, we give a description of the Virginia machine for this purpose. There is no patent on this machine, and any one may make it for himself. A cylinder (figure 7), as large as a flour barrel; is formed by [Pg 50] nailing narrow slats of plank, to two circular pieces of timber. The slats are put a little way apart, but not far enough for the pods to slip through when the cylinder is turned. A piece of timber runs lengthwise, through the centre of the cylinder, the ends of this project about a foot, and serve as an axle on which to turn it. A crank is attached to one end or both ends of the axle. Two pieces of scantling are fastened together in the shape of an X, one for each end, and these are held upright by having pieces nailed on horizontally, from one to the other. Several slats on the cylinder are fastened together to make a door, and this is attached to the cylinder by hinges, and fastened with a button.

The peanuts are poured into the cylinder, two or three bushels at a time, and it is made to revolve slowly, until all the earth and litter has fallen out. The door is then opened, the peanuts turned out and bagged.

In bagging the peanuts, care should be taken to have the sacks well filled. They are estimated to hold four bushels each, and if properly filled, good solid peanuts will over-run a little, especially in the first part of the season, before they are thoroughly cured. As the sacks are being sewed up, the corners must be packed with peanuts as long as any more can be got in. For sewing up the sacks, the planter needs a large peanut-sack needle and twine made purposely for this business. Sacks cost the farmer, at the present, ten cents each, and generally the peanuts are sold by gross weight and nothing paid for the sacks. In some markets the sacks are paid for, and a pound deducted from the gross weight, for each sack. If the planter sells to a merchant near home, he [Pg 51] seldom sews up the sacks, but ties them, and they are emptied and returned to him at the store.

Peanut "Factories." — It does not fall within our present plan to describe these establishments, any further than to give the reader, outside of the peanut belts, an idea of them. Formerly, many peanuts were sent into market without being properly assorted and cleaned, and it was found that, by assorting and re-cleaning them, a little margin of profit was left after paying expenses. One step led to another, and various appliances and machines were brought into requisition, until now, large buildings are devoted solely to the purpose of cleaning, assorting, and storing the peanuts. Some of these establishments employ many hands, both male and female, to clean, separate, and re-bag the peanuts ready for the trade.

Thus it has happened, that the business of cleaning peanuts has been taken out of the hands of the farmer, reduced to a system, and made a new industry. In fact, a division of labor; and now the merchant buys the peanuts of the planter just as they are picked, and the "factories," so-called, clean and assort them for the large buyers. Still, the merchant will pay more for Peanuts in nice order, and perhaps it would even now pay the farmer to properly clean and assort his crop before selling it.

The Best Markets.—A few years ago, the city of Norfolk was the sole market for the Virginia and North Carolina planter, and New York for the wholesale dealer. [Pg 52] Later on, Wilmington, Petersburg, Richmond, and several of the smaller towns began to buy peanuts, until now, every village and trading centre throughout the whole peanut belt, has become the repository for the crop of its own immediate section. Every year, the market has been coming nearer and nearer to the planter, until now he finds it about as profitable to sell to the nearest country merchant, as to ship to town, and sometimes more so. Frequently, the country merchant becomes the agent of some large buyer, who furnishes the capital, and he buys all the peanuts he can, at figures very near the ruling market price. Of course, this works very much to the planter's benefit. He sees his crop weighed, he escapes the middleman, with all the attendant expenses, such as commissions, freight, etc., he sells for cash, and he does not have to wait several weeks for returns.

Under this state of affairs, the home market, or home buyer, becomes the best for the farmer. And with the constantly increasing demand, and close competition between buyers, the cleaning factories are also coming nearer the farmer, and already exist, or will soon exist, in each of the counties and sections where the Peanut is much grown. Thus the planters generally, will soon be enabled to sell directly to the cleaners, and the latter to the wholesale buyers. So the planter will get market prices, without the trouble of going to market. Perhaps the competition will eventually grow sharper still, until, not only will the peanuts be cleaned and bought at home, but will also be manufactured into oil, flour, and the other commercial forms, in the sections where they are grown. In everything, the tendency now is, to carry [Pg 53] the factories to the raw material, and not the latter to the factories. It is not to be presumed that this crop will prove an exception.

Thus it is, that the farmer's work is being narrowed down, by the inevitable and beneficial law of the division of labor. The planter may now turn his attention wholly to the cultivation of the crop. How to order it, so as to realize the largest possible yield from the smallest possible areas, is now the problem before him. He finds given to his hands, a great and growing staple with great, and still unknown, possibilities, and he sees the demand becoming larger and more earnest, until now, the buyer comes to his very door, and puts down the ready cash for all of this crop that he has to sell.

Of course the planter must, and will bestir himself, to meet the ever-increasing demand. To do this with profit to himself, he must study this crop from beginning to end, he must learn the nature of the Peanut plant fully and correctly, and discovering how to increase the yield per acre to its maximum, unravel the secret of how to grow it at the least cost per bushel.

Picking Machines.—It may be well here to allude to a question, which, doubtless, the thoughtful reader has already asked himself, namely: Why does not some one invent a machine for picking peanuts rapidly, instead of having to do it by the slow and tedious process of hand-picking? In reply we state, that numerous attempts to do so have been made, but with very indifferent success. None of the many picking machines, that have hitherto been offered, have given satisfaction. It [Pg 54] seems that they cannot be made to do the work, and most planters appear to have given up looking for any help in this direction. Very recently, the writer has heard of one picking machine that is said to be giving satisfaction, but he has not seen it, or conversed with any one who has done so. That an efficient machine of this kind is an impossibility, is not believed, but whether anything can be made that would pay better than the old method, is the question. The planter must await developments. Perhaps some ingenious mechanic will take up the problem, and give the planter a perfect and cheap picking machine. Here is a field for ingenuity. A good machine would be a profitable invention. Who will try?

Having now traced the Peanut plant through the whole process of its planting, cultivation, harvesting, and marketing, the practical part of our task is ended. If the directions are such as will enable the beginner in this branch of rural industry, to successfully cultivate and manage this crop, the end will have been attained, and this little book will not have been written in vain. It has been prepared for those having no practical acquaintance with the cultivation of the peanut crop, not for the old and experienced planter. And yet, without egotism, it is believed that even the latter may find something in it that will be of use to him. Practices vary in different sections, even among men of the same calling, and inasmuch as methods herein detailed, will be found to vary from those practiced in North Carolina, Tennessee, Georgia, or the far South, so will the planter in [Pg 55] those States who may chance to read this treatise, be enabled to compare our methods with his, to see wherein they differ, and perchance may find here some point or plan a little better than his own.

It only remains now to give, in another chapter, some of the many uses of the Peanut.

CHAPTER VI.

USES.

Some of the more important uses of the Peanut and its plant are here given. In the course of time, as new discoveries are made, it is not improbable that the Peanut may subserve other valuable ends. But if no more uses than are now known, are ever found for any part of this plant, it will continue to occupy an important position among the agricultural productions of the country. Its importance will increase year by year, its value being too well understood and appreciated for it ever to lose its place among our leading crops.

Peanut Oil.—The use that gives the Peanut especial value as an American crop, is the place it occupies as an oil-producing plant. The oil of the Peanut is regarded as equal in all respects to sweet or olive oil, and may be employed for every purpose to which that is applied. This gives it at once a commanding position, and were no other use found for the plant, this would give it great importance among the economic productions of our [Pg 56] country. Olive oil is largely consumed for culinary uses, in medicine, and in the arts. Except in California, the olive has never been planted upon a commercial scale in this country, and it is very important that we possess a plant, that will obviate our dependence upon foreign oil. Of course, it is not within our scope to describe the manufacture of Peanut oil. The farmer is satisfied with knowing that his crops are in demand, and need not trouble himself about the methods by which they are converted into this or that useful commodity.

It is stated that a bushel of peanuts (twenty-two pounds in the hull) subjected to the hydraulic press, will yield one gallon of oil. The yield by cold pressure, is from forty to fifty per cent. of the shelled kernels, though if heat be used, a larger quantity of oil, but of inferior quality, is obtained. The best Peanut oil is nearly colorless, with a faint, agreeable odor, and a bland taste, resembling that of olive oil. It is more limpid than olive oil, and becomes thick when exposed to a temperature a few degrees below the freezing point of water. Peanut oil is not one of the drying oils. During the late war it was extensively employed in the Southern machine shops, and regarded as superior in its lubricating qualities to whale oil. For burning it is highly esteemed. The chief consumption of the oil is in making soap. For the production of oil for soap making, there were imported into Marseilles, France, from the West Coast of Africa, in one year, peanuts to the value of over five millions of dollars.

The residuum, or oil cake, may be sold for cattle feed.

Roasted Peanuts.—Almost every person residing in the eastern section of our country, must necessarily know [Pg 57] something of the value of roasted peanuts. One cannot pass along the streets of any of our larger cities and towns, without encountering, at every turn, the little peanut stands, where roasted peanuts are sold by the pint. They are kept for sale in numerous shops, they are peddled on the railroad cars, and sold to the loungers at every depot. Roasted peanuts are more common than roasted chestnuts once were, and almost everybody eats them. Even the ladies are fond of them, and frequently have them at their parties.

It is safe then to say, that everybody likes them, and finds them palatable, healthful, and fattening. From a pig to a school boy, no diet will fatten sooner than roasted peanuts. A person can live on them alone for an indefinite period, if eaten regularly and with moderation. The analysis of the Peanut shows it to be rich in the albuminoids, or flesh-forming elements. Roasted peanuts, therefore, form a very useful article of diet, and fill a place between the luxuries and the necessaries of common life. Wherever they have been once introduced, they cannot well be dispensed with; and as their use in this respect is constantly extending, this purpose alone would serve to keep the product before the public as a salable article. Once let the Peanut find its way to the great cities of Europe, and roasted peanuts be sold upon the streets there, as well as here, and the demand for them will far exceed the present limits, and the cultivation be necessarily extended over a much wider area than now. There is every reason to believe that the demand for the crop will continue to increase.

[Pg 58] **Peanut Candy.**—This is another of the purposes to which the Peanut has been applied, and serves to illustrate how varied and numerous are the uses of this remarkable production. Flat bars of sugar candy are stuck full of the broken kernels of the roasted nuts. It is quite good, and forms a pleasing addition to other kinds of confectionery.

Peanut Coffee.—Here again the Peanut fills a useful end, especially in times of scarcity, or high prices for coffee. Taken alone, and without any addition whatever of the pure berry, the Peanut makes a quite good and palatable beverage. It closely resembles chocolate in flavor, is milder and less stimulating than pure coffee, and considerably cheaper than Rio or Java. If mixed, half and half, with pure coffee before parching, and roasted and ground together, the same quantity will go as far and make about as good a beverage as the pure article, and a better one than much of the ground and adulterated coffee offered in the market. Indeed, if people will adulterate their coffee, it were much to be wished that they would use nothing more harmful than the Peanut for this purpose.

For making the beverage, the Peanut is parched and ground the same as coffee, the mode of decoction the same, and it is taken with cream and sugar, like the pure article.

Peanut Chocolate.—True chocolate is made by roasting and grinding to a paste, by the aid of heat, a very oily seed, the Cocoa-bean. In the preparation of [Pg 59] chocolate a great variety of articles are used to adulterate it and diminish its cost. Some of these, such as sugar and starchy substances, are harmless, while others, such as mineral coloring matters are injurious. Peanuts are largely used to adulterate chocolate, and so far as wholesomeness is concerned, are not objectionable. In containing a great deal of starch and oil, peanuts resemble the cocoa-bean, though without the nitrogenous principle, *theobromine* (which closely resembles *caffeine*), to which its nutritive qualities are largely due. Peanut chocolate is made in some Southern families by beating the properly roasted nuts in a mortar with sugar, and flavoring with cinnamon or vanilla as may be desired. Peanut chocolate, on so high an authority as the author, the late William Gilmore Simms, is vastly superior to peanut coffee.

Peanut Bread.—If peanuts are first mashed or ground into a pulp, and then worked into the dough in the process of kneading, no lard will be required to make good biscuit, and the bread will have an agreeable flavor, different from that imparted by lard, but of such a mild and pleasant taste as to be entirely unlike the peanut flavor. The skin of the kernel must first be removed, or it will impart a bitterish and nutty taste. There is some difficulty in doing this. Scalding does not do it very well. Strong soda water or lye, will quickly loosen it, so that it may be readily removed by rubbing with the hands, but either fluid would soon convert the Peanut into soap, and is, therefore, impracticable for this purpose. Could some cheap and handy machine be [Pg 60] invented, that would remove the skin from the kernel without loss, no doubt large quantities of peanuts would be used for bread-making purposes. Whether or not it would be economical, we cannot at present say.

Peanut Soap.—If a fair article of soap can be made of corn shucks, as was done in the South during the late war, then there can be no doubt that a better quality can be made from Peanuts. Surely a vegetable product containing such a large per-centage of oil, would be easily acted upon by lye. The writer has not experimented in this direction, but we hear of some who have tried it, and who say they have made a good and serviceable soap from the kernels of the Peanut without the addition of other oil or grease. We have no doubt but very good soap may be made from the Peanut, but whether the manufacture of such an article would be profitable at present prices, is another question. Perhaps for ordinary laundry soap it would not, but for the higher grades of toilet soap it might be. Here is a field for experiment, and yet we mention this use, as well as those of bread-making and coffee from the same article, as one of the possibilities of this plant, rather than a result to be looked for in the near future, if at all. It is well that manufacturers, and all others, should know what is capable of being done with this promising product. The more we can multiply the uses of any product of our farms, the wider will be the demand for it, and this is what the farmers desire.

Peanuts as Feed for Stock.—This is a use for the Peanut, about which we can speak with confidence, and from [Pg 61] experience. We now refer to the peanut pod, including, of course, the kernel, and not the vine or hay. Every kind of stock, horses, cows, sheep, hogs, and poultry, are exceedingly fond of the Peanut, and will leave any other food to partake of it. Cows, horses, and sheep eat the whole pod, hull and kernel together. Hogs and poultry (except turkeys) reject the hull, eating the kernel only. Turkeys, as a rule, swallow the pod whole, and a real live turkey can hide away quite a quantity of the nuts in a short time, if allowed free access to them. In fact, all animals do not seem to know when they have enough of this food. All stock fattens readily on them. The hog will lay on flesh faster on a diet of peanuts, than on corn, potatoes, or any other product with which the writer is acquainted. The poorest scrub of a hog, turned into a peanut field, after the crop is removed, and where he can get nothing but the pods he may find by rooting for them, will change his appearance in three days, and in a week, will be so much improved as hardly to be recognized as the same animal. As a pork producer we believe that the Peanut has not its superior in any clime or country. It is a thorough fat-former. Poultry intended for laying should be sparingly fed with it.

But we would not leave this subject without a grain of caution. While all stock fattens rapidly on the Peanut, it must be confessed that the fat is not always of the best quality. It is less firm and more oily than the fat derived from Indian corn, nor will the lard from hogs fattened upon peanuts show that pearly white and flaky appearance, which is the marked characteristic of pure lard made from corn. For this reason, most planters in [Pg 62] the peanut belt, feed their peanut-fed hogs on corn only for two or three weeks before killing them. This is done to make the lard firm and white, and in this manner, good pork and lard are produced at only a trifling cost. The hogs get nearly fat from the detached peanuts left in the field, and which otherwise would be lost. In this way the peanut-planter derives a very important benefit from this crop, apart from its value as a source of ready money. Were there no other use for the peanut, it would still pay well to raise it for making pork. In this case, the planting and cultivation would be the sole cost, as the animals would do all the harvesting. A very small field would fatten quite a number of hogs. Poultry intended for market, might well be fed on Peanuts, instead of corn or oats. The fowls would fatten faster and at less cost. In fact, we believe it would be economical to buy peanuts at ruling prices for fattening stock, especially old stock.

Peanut Hay.—If dug and cured before frost touches them, and before the leaves fall to any great extent, peanut vines make a very good provender for all stock. Some say it is better than blade fodder for horses and mules, but we are not prepared to advance this extravagant claim for it. It is, however, certainly an excellent article of fodder for cattle, sheep, mules, and horses, and if many sap peanuts are left on the vines, stock that is not worked much, will need no other feed during the winter months to keep them in good condition.

Most planters, accordingly, make it an object to try to save the vines for hay, and aim to dig the crop before [Pg 63] they are injured by frost. After a killing frost touches them, the vines are next to worthless as a feed. In fact, frost-bitten peanut vines are harmful, rather than beneficial, to stock, often causing colics, and endangering the life of a valuable horse or mule. Peanut vines, even the best of them, unharmed by frost, should not be fed very largely to horses. There is always a good deal of grit and dust upon them, and much of this taken into the stomach, cannot but be more or less harmful to the animals.

And yet, despite these few drawbacks, peanut hay has proved to be a valuable forage, and one that the peanut-planter could not well dispense with, inasmuch as so many do not make enough of other forage to serve them, and must, therefore, depend on the peanut crop to help them out. Thus the planter is benefited in several ways through this crop. He gets a valuable staple to sell, and one that always commands the ready cash, he fattens his hogs on the pods left in the ground, and he secures a large amount of very good hay in the vines. Thus he is doubly benefited, and no matter how low the price of peanuts may be, the farmer does not, and cannot, ordinarily, lose much on the cultivation of this great crop. If he does not risk too much on commercial fertilizers, which no planter of this crop ever should do, he runs little risk of suffering any crushing loss thereon.

Such is a brief but connected view of the Peanut crop from the time of planting the seed, to its sale and manufacture. The views and practice here advanced are all from original sources. We have not drawn upon any other writer for any part of this treatise. Indeed, save [Pg 64] a few short articles scattered through the agricultural press of the past ten or fifteen years, we know of no source from whence material could be derived. So far as we are aware, this is the pioneer work in America on the Peanut plant. This being the case, it must, of course, be quite defective. We might easily have made it a larger book, and perhaps some few years hence, when the field and subject shall have enlarged, it will be found desirable to revise and enlarge this treatise. For the present, we must be satisfied with smaller things, and remain content with a few practical directions rather than an elaborate work. Until that time, if it comes at all, we lay aside the pen, and turn our hands (as it has been our wont to do during the past few weeks) to actual labors in connection with the Peanut plant.

APPENDIX A.

STATISTICS.

It was our design, at first, to present a somewhat full array of statistics in relation to the Peanut. This, however, was soon found to be impracticable. The more we studied the few data at hand, the more were we convinced of their utter unreliability. The fact is, so far as the writer is aware, there are no credible data of this crop existing. No authoritative and systematic attempt to gather and compile the statistics of the Peanut has ever been made, and until this is done we shall never know its full extent and value. The "estimates" — mere guesses — of certain mercantile houses and newspapers, to express the bulk of the crop are, beyond a doubt, far wide of the mark. The following from a Georgia paper, is of this class:

"The goober [2] plays a more important part in commerce than might be supposed. We are all aware of its value as a social factor — of its influence upon oratory, music, and the drama — but how few of us know that one million nine hundred and seventy thousand bushels of this savory nut were consumed in this country during the twelve months ending on the thirtieth of September, 1883. These figures do not include the local consumption — say, for instance, in the rural districts of Georgia, where every substantial farmer has a patch of his own.

[Pg 66] "The figures relating to the goober crop make a column in the various prices current, but Georgia is not credited with any part of the crop. It seems that the goobers of commerce, so far as this country is concerned, are raised in North Carolina, Tennessee, and Virginia. In 1882, Virginia raised one million two hundred and fifty thousand bushels, Tennessee four hundred and sixty thousand, and North Carolina one hundred and forty thousand, making a total of one million eight hundred and fifty thousand. The aggregate value of the crop amounted to two million dollars. It is estimated that the peanut crop of 1883 will be at least two million bushels.

"We regret that Georgia has no place in these estimates. Goobers can be raised in this State as readily as in Virginia, and there is no reason why our farmers should not take advantage of the demand for them. The little patches for home use, could easily be increased to patches calculated to yield a comfortable supply of pocket money. As Georgians are known as goober-grabblers, there is no reason why they should not be known as goober-growers."

Still, these estimates serve a certain important end, and give an approximate idea of the magnitude of the crop. It is safe to say that it amounts to nearly three million bushels annually, and were all the information gathered that could be, it would doubtless be greater still. It is high time that the corps of statistical reporters to the National Department of Agriculture, were required to give the data for this crop, as well as for others, and some of them of less magnitude and value.

FOOTNOTE:

[2] See remarks on the term goober, in note on page 9.

APPENDIX B.

COSTS.

Perhaps the attentive reader has expressed surprise that so little has been said about the cost of planting, cultivating, and harvesting the peanut crop. This was because no estimate of costs that would suit one place, would apply in another and a distant locality. There is no uniformity in this matter, hence it was deemed best to leave each reader to count the costs for himself, based on his knowledge of his own local surroundings.

APPENDIX C.

THE PEANUT GARDEN OF AMERICA.

The following article from the Suffolk, Va., "Herald," gives a concise view of the growth and development of this staple in Virginia, and illustrates how a portion of the Southside has become, perhaps, the leading peanut-producing section of our country:

"When James H. Platt introduced his bill in Congress imposing a duty upon peanuts imported from Africa, a large majority of the members of that august body hardly knew what a peanut was. A few of them had eaten 'Goobers' which had been carefully cultivated in the garden by their grandmothers, but as to why they needed [Pg 68] protection, or how many of them there were to protect, but little was known even by the best informed. The culture of this important agricultural product was then in its infancy, and it was hardly recognized as an article of commerce.

"Only a few short years have rolled by, and what a change has been effected. The peanut crop has assumed gigantic proportions, and the aggregate amounts to millions of dollars, while the nut is in demand from one end of the Union to the other at satisfactory prices.

"The section of country contiguous to and lying south of James River, and between Norfolk and Petersburg, may be correctly termed the peanut garden of the world.

"In this section peanut farming has been brought to the highest state of perfection, and the average production per acre greatly increased from what was considered a good yield a few years ago.

"The one great difficulty in handling the crop seems to be, in the fact that no machine has yet been invented which will pick off the nuts from the vines in a satisfactory manner. This work must be done by hand, and as the entire crop matures at one and the same time, there is such a demand for labor during the picking off season that the supply is utterly inadequate to the demand. It is probable that within the next few years some plan will be devised for the successful storage of peas and vines until they can be conveniently picked off; and when this desirable end is accomplished, much of the rush and confusion incident to the gathering and marketing of the peanut crop will be avoided. This is already done by every thrifty planter who is able to hold his crop until [Pg 69] such time as he sees fit to sell it. He stores his peanuts away, and picks them off, mostly with his own force, at convenient intervals through the winter and spring.

"While so much has been done in the way of improvements in the production of the Peanut, those who have done the handling after reaching market have not been idle. In former years, only the bright shell and those well-filled, could be sold in the market. A dark color or half-filled pods was sufficient cause for rejection, and frequently they were on this account not even offered in market. Here, however, machinery was more successful. Various mechanical contrivances have been put in operation for cleaning and assorting the nuts, and to-day every grade of peanuts—from the large, plump, well-filled shell, to the smallest, blackest, and most insignificant half-filled pod—has a regular standard market value, according to the weight per bushel."

www.ingramcontent.com/pod-product-compliance
Lightning Source LLC
Chambersburg PA
CBHW030505220526
45464CB00006B/2665